SUPERWORLD

ELSA MEI GESSESSE

ISBN: E-Book # 978-1-970630-82-4
Paperback # 978-1-970630-83-1

Printed in United States of America

Disclaimer: The author and publisher make no representations or warranties with respect to the accuracy or completeness of the contents of this work and specifically disclaim all warranties, including without limitation warranties of fitness for a particular purpose. No warranty may be created or extended by sales or promotional materials. The advice and strategies contained herein may not be suitable for every situation. This work is sold with the understanding that the author and publisher are not engaged in rendering legal, accounting, or other professional services. If professional assistance is required, the services of a competent professional person should be sought.

Library of Congress Reg. # 2026907213

Cover Design by: Authors Hike

Publisher: Authors Hike

For permission requests, please contact: mcg0216@hotmail.com

About the Author

Elsa Gessesse
Email: info4book2025@gmail.cok

Website:
https://magazineymtv.my.canva.site/model-portfolio-elsa-gessesse

Instagram
https://www.instagram.com/elsa_missusa?igsh=Nzlkb3NzMW51M2Y5

Based in Las Vegas, Nevada, I am a multidisciplinary artist with professional experience in fashion, film, television, and live hosting. I am fluent in English and Mandarin and have received formal artistic training from an early age. I began studying ballet at age three and have trained in piano and violin since age five, building a strong foundation in performance and stage presence.

I am also a published author, having released three books by the age of eleven, with my work recognized on the Los Angeles Tribune bestseller list.

My other published books:

"The wolf who once loved"

" Just a game?"

In the fashion industry, I have participated in multicultural runway shows and International Beautiful You Fashion Tours since the age of six. I was invited to host the Dior Fashion Show in 2024 and the U.S. Fashion Show in Los Angeles in 2025. I have also held a title of Miss USA as well as Miss Nevada. further strengthening my experience in pageantry and high-profile events.

My acting credits include appearances in a music video and multiple international commercials, including campaigns for JW Marriott International. I have played significant roles in two Hollywood feature films and was awarded Most Influential Actor of the Year at the Hollywood Golden Feather Awards in 2025.

In addition to film and fashion work, I have extensive experience as a bilingual host for major international events. I served as a host for the CCTV Children's New Year Gala in both 2024 and 2025, and I was a lead performer and host at the 2024 Youth China International Closing Ceremony in Beijing. I also hosted the CCTV Media Center National Awards in Beijing in 2023.

I bring a global perspective to my work, having traveled to fifteen countries. My international experience, performance training, and versatility across fashion and acting allow me to adapt seamlessly to diverse creative environments.

Acknowledgment

No world is built alone.

This story began as a question and grew into a universe through the patience, curiosity, and belief of many people. I am deeply grateful to my family, whose steady support anchored me while my imagination drifted across distant stars and unfamiliar futures. Your faith in me made the long journey possible.

To my friends and early readers, thank you for challenging the ideas, questioning the science, and helping the story find its shape. Your feedback strengthened both the logic of the world and the heart beating at its center.

I owe special thanks to the writers, scientists, and dreamers—past and present—who dared to imagine what lies beyond the known. This book stands on the shoulders of their curiosity.

And finally, to the reader: thank you for stepping into this universe. May these pages spark wonder, ask difficult questions, and remind us that the future, like fiction itself, is something we choose to create.

Table of Contents

About the Author i

Acknowledgment iii

Chapter One The Girl Who Fixed Everything 1

Chapter Two The First Cracks 5

Chapter Three Questions in the Perfect Day 7

Chapter Four The First Cracks 9

Chapter Five Whispers in the Songs 12

Chapter Six The First Key 13

Chapter Seven A Question Asked 15

Chapter Eight Small Warnings 17

Chapter Nine The Key, And Doubt 18

Chapter Ten The First Unease 20

Chapter Eleven The Second Key 21

Chapter Twelve Testing the Rules 23

Chapter Thirteen Shadows of Doubt 25

Chapter Fourteen Normal Lucy 27

Chapter Fifteen Fighting Old Instincts 28

Chapter Sixteen Doubt Creeps In 30

Chapter Seventeen Lucy is Ordinary 32

Chapter Eighteen Following the Rules 33

Chapter Nineteen Doubt Creeps In 35

Chapter Twenty Avoiding Temptation 36

Chapter Twenty-One Following the Rules 38
Chapter Twenty-Two A Glimpse Beyond 39
Chapter Twenty-Three The Crush Lurks 41
Chapter Twenty-Four First Glimpse 42
Chapter Twenty-Five Danger! 44
Chapter Twenty-Six False Safety 46
Chapter Twenty-Seven The Crush's Shadow 48
Chapter Twenty-Eight The Final Rush 49
Chapter Twenty-Nine Gone 51

Chapter One
The Girl Who Fixed Everything

Bad days had patterns.

Bea knew hers by heart. The bus was always late when she needed it most. The coffee machine at work sputtered like it was offended by her existence. Her phone stayed silent for hours and then buzzed all at once, delivering bad news in neat little notifications.

It wasn't dramatic. It was just… constant.

So when Bea stepped outside that morning and the air smelled sweet—like clean fabric and something warm—she stopped.

She stood there for a second too long, keys dangling from her fingers, staring at the street like it might explain itself.

Nothing looked different. Same cracked sidewalk. Same crooked stop sign. Same bus bench with peeling paint and a gum stain shaped like a star.

Except for the girl sitting on it.

She looked out of place in a way Bea couldn't explain. Not flashy. Not strange. Just… precise. Like she'd been placed there intentionally.

The girl wore a pale yellow coat, buttoned neatly, her legs crossed as she hummed under her breath. The sound was soft, almost blending into the hum of the city. When she looked up, she smiled.

It wasn't wide or forced. It landed exactly right.

"You're Bea," the girl said.

Bea's stomach dipped. "Sorry—do I know you?"

The girl tilted her head slightly, considering. "Not yet."

That should have set off alarms. Instead, Bea felt her shoulders loosen, like she'd been holding her breath all morning without realizing it.

"I'm Lucy," the girl said, standing. "You look like you could use a better day."

Bea huffed a laugh before she could stop herself. "That's not really something you can promise."

Lucy's smile didn't fade. "I know."

The bus arrived early. Bea noticed because she almost missed it.

Lucy stepped on beside her, tapping her card with perfect timing. They sat together without discussing it, the seat warm and surprisingly comfortable. The bus ride felt… quiet. Not awkward-quiet. Peaceful.

Bea checked her phone out of habit.

No missed calls. No reminders. No buzzing anxiety pressing against her ribs.

Lucy noticed. "Sometimes things get quieter," she said.

Bea raised an eyebrow. "That's not normal."

Lucy shrugged lightly. "Normal's overrated."

They got off two stops early. Bea didn't remember deciding that.

The café on the corner smelled like vanilla and steamed milk. It was crowded, but not overwhelming. That's where they met Erin.

Erin bumped straight into Bea, iced coffee sloshing over the rim and onto Bea's sleeve.

"Oh my god—I'm so sorry," Erin blurted, already panicking. "I swear this only happens when I'm already late—"

Lucy was there instantly, handing over napkins.

"It'll wash out," Lucy said gently.

Erin blinked. "You're… really calm."

Lucy smiled. "It helps."

The stain faded before Bea's eyes. Not slowly. Not faintly. Just—gone.

"That's new," Bea muttered.

They stayed. One coffee turned into two. Erin stopped checking the time. Her phone rang halfway through their conversation, and she stared at it like it might disappear.

"They called me back," Erin whispered. "They never do that."

Lucy didn't look surprised.

Mina joined them next, sliding into the empty chair with a relieved sigh. She'd been waiting on a decision, a message, something that would tell her which way her life was going. Five minutes after Lucy arrived, her answer came.

Positive.

Theo found them laughing.

He said he'd just been walking by and felt like stopping. He had that kind of warmth—easy, grounding. By the end of the afternoon, it felt like they'd all known each other longer than a day.

Bea noticed something strange as the hours passed.

Nothing went wrong.

No spilled drinks. No missed calls. No awkward silences. Even the sky stayed clear, a soft blue stretched thin overhead.

It was… nice.

That night, Bea lay in bed staring at the ceiling, waiting for the familiar weight to settle back into her chest.

It didn't.

The days after that blurred together, sweet and gentle. Lucy always showed up at the right moment. When Bea's alarm didn't go off, Lucy was already knocking. When Erin doubted herself, Lucy knew exactly what to say. When Mina hesitated, Lucy smiled and things worked out anyway.

"It's just luck," Lucy would say, laughing it off.

Bea started to believe her.

Life felt lighter. Brighter. Like someone had finally fixed a loose wire in the world.

It wasn't until Erin mentioned it—quietly, almost joking—that Bea felt the first crack.

"Doesn't it feel… too easy?" Erin asked one evening.

Bea opened her mouth to answer.

And for the first time since meeting Lucy, she hesitated.

Chapter Two
The First Cracks

The next morning, Bea woke to the smell of pancakes. Sweet, fluffy, perfect. She rubbed her eyes, expecting her apartment to be messy or her kitchen disaster-prone as usual. But no—everything was in place. Plates stacked neatly. A fork beside each pancake. Syrup in a small dish, not sticky on the counter.

Lucy was humming in the corner, flipping a pancake with exact precision.

"Good morning," she said. "Sleep well?"

Bea stared. "Yeah… I guess. How do you do this?"

Lucy just smiled, almost teasing. "Do what?"

"Make… everything… perfect."

Lucy shrugged. "Some people notice, some don't. You notice. That's good."

Bea felt an odd warmth spread through her chest. Maybe it was relief. Maybe it was… something else.

By the afternoon, Erin, Mina, and Theo joined them. Their days had followed the same pattern: minor disasters solved before they even had time to react. Phones that wouldn't ring when they shouldn't, emails sent at the exact right moment, errands that ran smoothly.

It felt like magic.

And then the first song played.

It came from the tiny radio in the café corner, soft and almost drowned out by chatter:

"Follow the path you think is true,

The one who shines will guide you through..."

Bea shivered. Something in the melody tugged at her memory. She'd never heard it before, but the lyrics felt... personal.

"Did you hear that?" Mina asked, glancing at the radio.

"It's just a song," Theo said, shrugging.

Lucy tilted her head. "Songs don't choose who listens. They just find you."

Bea frowned. Something about that didn't sound comforting at all.

By the end of the week, they had discovered more songs. Each one seemed tailored—telling stories of love, luck, and a mysterious "super world." Bea noticed patterns: every song hinted at someone helping someone else, someone guiding the lost. The words were innocent... almost too innocent.

Erin was the first to speak up. "Do you ever... feel like this is too perfect?"

Bea hesitated, glancing at Lucy. She was arranging sugar packets in neat little lines, humming again. Everything about her was calm. Too calm.

"Yes," Bea said finally. "It feels... weird."

Lucy glanced at her, eyes soft. "Weird can be good. Weird can be... protection."

Bea tried to nod, but the pit in her stomach only grew.

Chapter Three
Questions in the Perfect Day

By Monday, they had formed routines. Bea walked to school with Lucy. Erin joined for lunch. Mina and Theo rotated their afternoons with study sessions and walks. Everything followed a rhythm: pleasant, safe, predictable.

But the songs kept coming.

One afternoon, while Bea was walking home, a soft tune floated through her headphones.

"The one you follow is not who you think,

But who gave you the wings you never had…"

Bea froze. Her heart thumped. Wings?

She tugged the headphones off, looking around. Nobody was nearby. Nothing had changed. But the words… felt pointed.

When she told Erin and Mina later, they exchanged uneasy glances.

"Do you think it's… about Lucy?" Mina asked cautiously.

Bea swallowed. "I don't know. I hope not. But…"

Lucy, as if she had read Bea's thoughts, appeared behind them suddenly. "About me?" she asked softly.

Bea blinked. "Uh… maybe?"

Lucy smiled, a little too perfectly. "It's easier to enjoy the ride when you trust who's guiding you."

Erin frowned. "Guiding us where?"

Lucy shrugged lightly. "Somewhere better. Somewhere safe."

Bea wanted to ask more, and wanted to press. But the words stuck in her throat. Something in Lucy's calmness warned her: questions were optional. Answers… were not guaranteed.

And the songs kept playing.

Chapter Four
The First Cracks

The morning air smelled like warm bread and something metallic. Bea frowned. Nothing in her apartment had changed, but somehow it didn't feel right.

She checked her phone. No notifications. No missed calls. Her day should have been normal.

But the first odd thing happened outside: a stray cat ran in front of her. Its eyes were too sharp, too aware, and it stared at her long enough that her skin prickled. When she blinked, it was gone.

Lucy was waiting at the corner, humming softly, hair catching the light. She smiled. "Good morning."

Bea hesitated. "Lucy… do you ever feel… like things are… too perfect?"

Lucy tilted her head, her smile steady. "Perfect? I don't think I've ever heard that word sound like a complaint."

"It's just… the songs," Bea said. "They… they feel personal. Like they're speaking to us, not just… randomly playing."

Lucy's lips curved just slightly. "Songs don't play randomly, Bea. Some songs choose who listens."

Bea shivered, something tightening in her chest.

At school, Erin, Mina, and Theo all noticed the same strange coincidences: tests they hadn't studied for came out easier than expected, a teacher gave them extra time at just the right moment, a lost notebook appeared in the perfect spot.

"It's like someone's… guiding us," Erin whispered.

“Guiding us… or controlling us,” Mina muttered, and everyone froze.

Bea didn’t want to admit she had started thinking the same thing. But a tiny voice in her head whispered: what if we’re not lucky? What if we’re trapped?

That evening, they gathered at Lucy’s apartment—well, the apartment she had “perfectly arranged,” but Bea barely noticed the furniture anymore. Her eyes kept going back to the little radio on the shelf.

It crackled, and then a soft voice sang:

“Follow the hand that lifts you high,

The one who made your shadows fly,

Hearts that trust will reach the sky,

But lose your way, and say goodbye.”

Bea’s stomach dropped.

“Lucy,” Erin said, frowning. “The lyrics… they’re talking about someone. Some… guy?”

Lucy laughed softly, too soft. “They talk about whoever matters most to you.”

“Who matters most to us?” Mina asked, tilting her head. “I thought… I thought we just… lucked out?”

Lucy hummed, turning a knob on the radio. The next song started immediately, melody bright and sweet:

“Chase the light you think is true,

The wings you wear were given to you,

Don’t look back, the doors are few,

Keep the guide in sight, see it through.”

Bea felt her pulse spike. The words… they felt like instructions.

"Guide?" Theo asked. "What guide?"

Lucy's eyes softened, but something in her smile didn't reach the corners. "The one who brought you here."

Bea's throat tightened. "Brought… us… where?"

Lucy hummed again, and the radio fell silent. "Some places you'll only understand when it's too late."

The room felt colder. Everything felt sharper.

And for the first time, Bea noticed something she hadn't before: Lucy's shadow didn't always match her movements.

Chapter Five
Whispers in the Songs

The next morning, Bea woke to the same sweet smell of breakfast. Pancakes. Toast. Coffee. Everything exactly perfect. She blinked. It was comforting… almost too comforting.

Erin came rushing in, holding her phone like it was a fragile treasure.

"You guys have to hear this," she said, pressing play on a small speaker.

A new song started, soft and melodic:

"The path is shown, but not by light,

The one you follow may hide from sight,

Trust the hands that lift you high,

But watch them closely, do not comply."

Bea shivered. Something about that "do not comply" line made her stomach knot.

"Sounds… weird," Theo said, frowning. "It's almost… like a warning."

Lucy was nearby, arranging napkins on the table. She hummed the same tune along with the song, soft and light. "Warnings are for people who don't listen," she said with a smile.

Bea noticed for the first time that Lucy's humming didn't quite match the music. Her voice was a little off, a little ahead, almost like it was guiding the song instead of following it.

Chapter Six
The First Key

Later that day, Lucy took Bea, Erin, Mina, and Theo to a small, dusty bookstore tucked between two cafés.

"There's something I want to show you," Lucy said.

Bea didn't realize how quickly she was following until she was holding a small, brass key. It was warm, heavy in her hand, and engraved with strange symbols she didn't recognize.

"This," Lucy said softly, "is the first key. It opens doors you can't see yet."

Bea's chest tightened. "Doors… to what?"

Lucy smiled that too-perfect smile again. "To your better life. Don't worry… it's all part of the songs you've been hearing."

Erin glanced at the key, uneasy. "It feels… important. But I don't know why."

Lucy hummed. "Sometimes importance finds you before understanding does."

Bea put the key in her pocket, a small shiver running through her. Something about it felt… alive.

Over the next few days, small things began to feel off, Mina noticed shadows that didn't match the light. Theo heard whispers in empty rooms. Erin felt like the songs were repeating hints only she could hear.

Bea tried to ignore it, focusing on schoolwork and friends. But when she glanced at Lucy, something about her calm smile made her heart pound.

One evening, as they sat together at a café, the radio started another song:

“The guide is close, yet far from view,

The path you take is shaped by who,

Looks like light may hide the truth,

And doors you seek may not be new.”

Bea’s pulse raced. Something in the words felt urgent.

Lucy hummed along again, but her eyes were distant, unreadable.

Chapter Seven
A Question Asked

Bea stayed up that night staring at the ceiling.

"What are we really following?" she whispered to herself.

She thought about Lucy, the keys, the songs. Everything that had been perfect suddenly felt fragile.

When she shared her worries with Erin the next morning, Erin didn't laugh. She just nodded.

"I've been feeling it too," Erin said. "Like… maybe Lucy isn't just helping us. Maybe she's… controlling more than she lets on."

Bea shivered. "I don't want to believe that…"

"Me neither," Erin admitted. "But the songs… they're not normal. They're guiding us. And the keys… something about them isn't right."

Bea nodded slowly. Something deep in her gut whispered: we're in deeper than we realize.

The following week, Lucy announced a small trip for the group.

"Don't worry," Lucy said. "Just a little adventure. You'll like it."

Bea felt unease coiling in her stomach. Something about the way Lucy said "you'll like it" didn't sound optional.

They followed her to a small, hidden park at the edge of the city. The trees swayed in a gentle wind. A fountain bubbled in the center. Everything was… perfect. Too perfect.

And then the radio played again, faint at first:

"Beyond the world you know,

Lies a place the chosen go,

Perfect days may fade to gray,

But doors you find may show the way."

Bea stared at Lucy. Her smile didn't falter.

Something about the words… they weren't just music. They were instructions.

And for the first time, Bea wondered if Lucy was leading them somewhere they couldn't leave.

Chapter Eight
Small Warnings

Bea woke to a dull buzzing in the morning. Her alarm didn't go off. She frowned, but the city outside looked bright, soft sunlight spilling over the rooftops. Lucy was waiting outside, humming a quiet tune.

"You're late," Lucy said softly.

Bea blinked. "I… my alarm didn't—"

Lucy smiled, tilting her head. "Sometimes things work out anyway."

Her voice sounded innocent, but Bea felt a tiny spike of unease. Too perfect. Too controlling.

The friends noticed patterns. Every test, every quiz, somehow came out easier than expected. Tasks they forgot to do were suddenly completed. Misplaced items appeared at exactly the right moment.

Bea whispered to Erin, "It's like… we're living in someone else's schedule."

Erin's eyes widened. "I thought it was luck. But… it's too consistent."

Lucy was nearby, rearranging sugar packets again, humming a melody they hadn't heard before. Every note seemed intentional.

Bea shivered. Something about Lucy's calmness made her heart pound.

Chapter Nine
The Key, And Doubt

Lucy pulled them aside one afternoon.

"There's another key," she said, sliding a small silver key across the table. Its surface was engraved with symbols Bea didn't recognize.

Bea picked it up. It was heavier than it looked, warm to the touch.

"What doors does this open?" she asked.

Lucy smiled softly. "The ones that matter most."

Erin frowned. "Why do we need keys for doors we can't see?"

Lucy tilted her head. "Sometimes, seeing isn't as important as following."

That night, Bea heard the faint radio song drifting under her bedroom door:

"A world unseen, yet near to hand,

Follow closely, understand,

Doors appear for those who try,

But do not stray, lest it pass by."

She froze. It felt… like it was speaking to her directly.

Mina leaned over from the couch. "Do you hear that too?"

Bea nodded. "It's… too personal."

Theo frowned. "Songs don't talk like that. It's… instructions."

Bea swallowed. Her stomach twisted. Lucy's humming joined the melody, soft and guiding. She realized something terrifying: Lucy's voice didn't just follow the song. It wasn't normal…

Bea couldn't sleep. She stared at the ceiling, thinking about Lucy, the keys, the songs.

Erin whispered from her bed. "Do you think… maybe Lucy's not just helping us?"

Bea's eyes widened. "I don't want to think that…"

"But I do," Erin said. "It feels… like she's controlling more than she lets on."

Bea nodded slowly. The voice in her head whispered: we're in deeper than we realize.

And for the first time, the idea that the perfect life Lucy gave them might not be safe… didn't seem impossible.

Chapter Ten
The First Unease

Bea had stopped noticing the small coincidences at first. The perfectly timed messages, the lost items appearing exactly where they were needed, the flawless breakfasts—she had begun to take them for granted. But that morning, something caught her eye.

The sunlight fell through her window in stripes, perfectly parallel, and for a moment she noticed that Lucy's shadow didn't line up with hers or the furniture. It was almost imperceptibly shifted to the left, and Bea's stomach twisted. She shook her head, telling herself she was imagining things, but the unease clung to her like a second skin.

At breakfast, Lucy hummed a tune Bea hadn't heard before, light and lilting. It seemed harmless, but as Bea glanced around, she noticed that Erin, Mina, and Theo all paused mid-action, waiting for Lucy to finish. Their hesitation was subtle—too subtle. A tension lingered in the air, as if they were all holding their breath, waiting for some invisible signal.

Bea tried to speak up. "Does anyone else feel… weird today?"

Mina looked up from her toast, a faint frown creasing her brow. "I don't know… a little, maybe?"

Lucy's head tilted, and her smile was calm, soft, perfect—but her eyes didn't meet Bea's. "Weird can be the start of discovery," she said. "Or a warning. It depends on how you listen."

Bea swallowed. The words were comforting, almost, but a tiny part of her twisted at the sound of them. She realized that maybe this "perfect life" had cracks she hadn't noticed before.

Chapter Eleven
The Second Key

Lucy produced a small silver key that afternoon, placing it in Bea's hands as though it had been waiting for her. The metal was cold at first, then warmed to her touch. Symbols were etched into it, intricate and deliberate, and Bea could feel the weight of their meaning pressing into her fingers.

"This opens doors," Lucy said softly, "doors that most people don't even realize exist. But they matter. They matter more than you think."

Bea stared at the key. "What doors? And… why me?"

Lucy's smile was calm, almost gentle. "Because you're ready. Or at least, I think you are. The doors will find you when you need them."

Erin frowned, hesitating before speaking. "But… how can we follow doors we can't see? How can keys work if we don't know what they open?"

Lucy hummed, a note that seemed to vibrate just enough to make the room feel alive. "Seeing isn't always necessary. Following is."

Bea's chest tightened. The warmth of the key in her hand seemed almost alive. She wanted to drop it, but some force, some instinct, made her clutch it closer.

That night, she lay awake, staring at the ceiling. For the first time, she wondered if their perfect life wasn't perfect at all, but a carefully designed path someone else had made.

The radio, faint and nearly drowned out by the chatter of the café, played another song. This one sounded brighter than the others, almost cheerful. But the lyrics cut through Bea like a whisper in the dark:

"The guide is near, yet far from view,
The hands you trust may lead you too,
Paths converge where hearts align,
And doors appear when stars combine."

Bea froze, fingers tightening around her cup.

"They're… telling us something," she whispered.

Theo leaned forward, voice low. "It's not a song. It's a map. Instructions."

Lucy hummed along, shaping the melody as it played, almost imperceptibly altering the pauses and emphasis. Bea felt her stomach churn. Her mind screamed that the songs were more than music. They were part of Lucy's design, and Lucy's soft voice confirmed it.

Mina shivered. "Do you think… maybe we're being tested?"

Bea shook her head slowly. "I… I don't know. But it doesn't feel like a test I signed up for."

And for the first time, Bea realized how much power Lucy had, not just over events, but over the way they thought. Every coincidence, every perfectly timed message, even the songs. They were all threads in a web, and Lucy was weaving it quietly, invisibly.

Chapter Twelve
Testing the Rules

By the weekend, Bea and the others began testing boundaries. They tried small acts of rebellion: walking different routes home, leaving the pancake breakfast untouched, ignoring Lucy's suggestions. At first, nothing happened. But the more they resisted, the more life nudged them back. The bus they didn't take arrived early. Messages they ignored still reached them at exactly the right time. Lost items they had ignored appeared in the perfect spot.

Erin whispered, her voice tense, "It's like… we can't make mistakes. Like the world itself is bending around us."

Bea's hands shook slightly as she held the second key. "It feels like a trap," she said quietly. "Like we're living someone else's story, and we don't even realize it yet."

Lucy appeared at that moment, calm, her smile gentle but her eyes sharp. "The story is yours to live," she said softly. "You just haven't understood it fully yet."

Bea realized with a shiver that Lucy's presence itself was part of the rules. Her calmness, her smile, the way she hummed the songs, it all guided them, corrected them, and constrained them.

The perfect life they had thought they were enjoying was actually a carefully maintained pattern.

Lucy suggested a small trip for the group, claiming it would help them understand the songs and keys better. Bea felt a prickle of unease in her stomach. Lucy's phrasing, "you'll like it," didn't feel optional.

They arrived at a secluded park on the edge of the city. Every tree, flower, and even the fountain was perfectly arranged. Nothing seemed accidental. Everything was curated to an almost impossible degree.

The radio played again, faint beneath the soft rustle of the wind:

"Beyond the world you know,
Lies a place the chosen go,
Perfect days may fade to gray,
But doors you find may show the way."

Bea felt the words like a chill sliding down her spine. They weren't just songs anymore. They were instructions, warnings, and maps, all layered into one.

Lucy's smile was calm, almost maternal, but her eyes hinted at something else. A depth of knowledge, a plan, something Bea couldn't yet grasp.

Standing there, Bea realized with sudden clarity that they might already be inside something far bigger than themselves. A place they could not leave freely. A "Superworld" orchestrated by Lucy, slowly revealing itself piece by piece.

And for the first time, Bea truly understood that the stakes were far higher than she had ever imagined.

Chapter Thirteen
Shadows of Doubt

The morning was quiet, almost too quiet. Bea lingered by the window, watching the city move with a rhythm that suddenly felt foreign. Lucy sat across the table, flipping pancakes with an easy, normal smile. She laughed at a joke Mina made about spilled coffee the day before, perfectly casual.

Bea felt a strange tension in her chest. Yesterday, the key had felt alive. Today, Lucy was just… ordinary. No humming, no guiding presence, no hints of hidden control. She was just… Lucy. And somehow, that was more unsettling than anything Bea had seen before.

"Maybe… maybe we're overthinking," Erin muttered, reaching for her tea. "The songs… maybe they're just… songs. Maybe nothing is trying to control us."

Bea frowned. "But the keys… the patterns… how everything just works out perfectly?"

Lucy caught her gaze, smiling softly. "Sometimes life just… works out. Maybe it's luck. Maybe it's a coincidence."

For the first time, Bea wondered if she had been projecting all the fear onto Lucy. Maybe the calm, perfect life wasn't dangerous. Maybe it was safe. Maybe… maybe the songs were warning about someone else entirely.

That evening, Mina pulled Bea and Erin aside, her voice barely above a whisper.

"I… I think I get it," she said. "The songs… the lyrics about the 'crush'… maybe they aren't about Lucy at all. Maybe they mean…

someone else. The one they're all singing about, the one we're supposed to... stay away from."

Bea's eyes widened. "You mean... the guy? The one in the songs?"

Mina nodded. "Yeah. That's why all the lyrics mention chasing attention, getting noticed, and trying to make him happy. Maybe Lucy isn't controlling us at all. Maybe she's... normal."

Erin's brows furrowed. "So all this time... all the warnings we thought were about Lucy... they're about... crushes? About not falling for someone?"

Bea swallowed. The thought made her chest tighten. It fit. Too well. Lucy had been acting so normal lately, almost human. Too human. Could they have been blaming her for something she never did?

Chapter Fourteen
Normal Lucy

The next few days, Lucy's presence became almost mundane. She laughed, told stories about her childhood, brewed tea for everyone, and helped Theo with his homework. There was no guiding hum, no subtle shaping of events.

Bea found herself relaxing slightly, though uneasily. The patterns were still there, things working out, keys appearing, but maybe Lucy wasn't orchestrating them. Maybe it had nothing to do with her at all.

The songs, however, continued. Their lyrics were persistent:

"He waits beyond the world you know,
Keep your heart guarded, don't let him show,
Doors appear if you don't give in,
Perfect worlds fade if love slips in."

Every line made Bea's stomach twist. They were clear and urgent, yet no longer pointing to Lucy. The "he" in the lyrics, the crush, suddenly seemed like the real threat.

Chapter Fifteen
Fighting Old Instincts

Bea, Erin, Mina, and Theo began testing the theory. They avoided any thoughts about the crushes hinted at in the songs, refrained from seeking attention, and monitored their own behavior.

It was harder than it seemed. The world seemed to tempt them with coincidences, small chances to glance at messages, small invitations to think about him. Each time they almost faltered, the radio played softly in the background, a faint melody nudging them back.

Lucy noticed none of it. She was ordinary, almost invisible in her perfection. No manipulations, no subtle commands. Just… normal.

Bea caught herself thinking, maybe I've been wrong all along. Maybe Lucy isn't the danger… maybe the songs were right about him all along.

The shift in perspective changed everything. Suspicion between the friends was subtle but real. Each song now felt like a puzzle to solve, not who to trust, but who to avoid.

"The chosen hearts must guard their path,
Avoid the eyes that spark the wrath,
Doors appear for those who wait,
Love too soon could seal your fate."

Bea felt the lyrics pressing against her thoughts. Avoid the eyes. Avoid the crush. She didn't know if it was metaphorical or literal, but the urgency in the melody made her pulse quicken.

Lucy, ordinary and harmless, poured herself a cup of tea and laughed at Mina's story about the library. She looked like she belonged in the world Bea had always imagined, safe, human, nothing supernatural.

And that terrified Bea. Because if Lucy wasn't the danger, then the songs were right. Someone else was.

Someone was waiting beyond the world they knew.

Chapter Sixteen
Doubt Creeps In

The morning sunlight spilled into the apartment, brushing across the floor in warm streaks. Lucy was humming while she stacked dishes in the kitchen, perfectly ordinary, perfectly human. Bea paused mid step, watching her. The fear she had carried for weeks suddenly felt misplaced. Lucy wasn't manipulating anything. She wasn't guiding them with hidden rules or subtle commands. She was normal.

Mina, Erin, Theo, and Bea gathered around the kitchen table, all noticeably tense. The radio in the corner crackled and sputtered, then played a soft melody that seemed to float around them like a warning:

"He waits beyond the world you know,
Keep your heart guarded, don't let him show,
Doors appear if you don't give in,
Perfect worlds fade if love slips in."

Bea's stomach twisted. The "he" in the lyrics suddenly seemed more real, more threatening than anything Lucy had ever done. She glanced at her friends, and they all looked uneasy, as if realizing the same thing at once.

"The crush," Mina whispered. "It's him. Not Lucy."

Erin nodded slowly. "All this time maybe Lucy was never dangerous. The songs are warnings about someone else. About not falling in love, not letting someone pull us into danger."

Bea shivered. Her hands shook slightly. "So everything we feared about Lucy maybe it was never true. She's safe. She's human. But he, he's the threat."

The radio hummed again, almost like a sigh, as if confirming their realization.

The next few days became exercises in discipline. Every glance at the school hall, every passing smile, every faint thought of attraction was carefully monitored and suppressed. The friends didn't just avoid crushes. They trained themselves to ignore even the smallest sparks of interest.

Bea noticed herself watching her own thoughts, her feelings, her impulses. Every time she felt drawn to someone, even in the tiniest way, a sharp mental alarm went off. The songs repeated in her mind, instructions disguised as melodies:

"Eyes that linger, hearts that stray,
Doors appear to those who stay,
Trust the hand that gives no harm,
But keep away from the subtle charm."

"Step by step, doors are shown. Perfect worlds fade if love slips in," she whispered to herself. The lyrics pressed against her mind, a gentle but insistent warning she couldn't ignore.

Erin watched Bea carefully. "We have to follow the songs. Avoid the crush, ignore temptation. It's the only way."

Bea nodded, though her heart ached at the thought. They were safe from Lucy, but not safe from the rules, the songs, or the danger lurking just beyond the edges of the world they knew.

Chapter Seventeen
Lucy is Ordinary

Lucy continued to act entirely normal. She baked muffins for everyone one morning, laughing at Theo's clumsy attempt to juggle his utensils. She helped Mina organize her notes and even teased Erin gently about spilling coffee the day before.

Bea observed her quietly, half expecting some hidden manipulation, some flicker of control. But there was nothing. Lucy was human. Ordinary. Innocent in the way she moved, spoke, and smiled.

For the first time in weeks, Bea felt a flicker of relief. But it didn't last. The world around them, the perfect coincidences, the keys appearing just when needed, the radio songs, still hummed with subtle insistence. Someone, or something, was orchestrating events. And it wasn't Lucy.

Mina glanced at Bea with a frown. "It's strange. We've spent so long fearing Lucy, but maybe we've been looking in the wrong place."

Bea nodded slowly. "Yes. The danger isn't Lucy. It's the crush, the rules, the Superworld itself. Lucy is part of our ordinary world. She's not controlling us."

Chapter Eighteen
Following the Rules

The friends focused on the songs and the keys, treating them as a code to follow. They memorized lyrics, watched coincidences carefully, avoided any signs of temptation, and refrained from thinking about the crushes mentioned in the music.

"Hold your heart and guard your eyes,
The chosen paths are safe from lies,
Step by step, the door is shown,
But love misplaced will lead alone."

The rules were clear. Each key, each song, each perfectly timed event was part of a pattern they couldn't yet fully understand, but they could follow it.

The tension in the group increased, though. Avoiding natural impulses and monitoring their own feelings constantly was exhausting. They all felt it. The safe path was narrow, and one misstep could unravel everything.

Bea held the second key tightly. She could feel the faint warmth in her palm, like it knew she was ready for the next step, the next door. But something still held her back.

One evening, as the sun dipped behind the city, Bea noticed the air felt different. The familiar scents, the bread, the coffee, even Lucy's perfume, shifted. Shadows grew longer, sharper at the edges, and the world seemed to stretch in subtle, unnatural ways.

The radio whispered another melody:

"Beyond the world you know,
Lies a place the chosen go,

Perfect days may fade to gray,
But doors you find may show the way."

Bea's stomach clenched. The Superworld was no longer distant. It was near, curling around their lives with invisible fingers, waiting for them to misstep.

Lucy, ordinary and harmless, laughed at a joke Mina told. She poured herself a cup of tea, completely unaware of the tension building around her. She was safe. Perfectly safe.

And that terrified Bea. Because if Lucy wasn't the danger, then the Superworld and the "crush" lurking beyond the edges of their lives were.

Bea realized with cold clarity that they were already approaching the edge of something vast, something beyond their understanding, and the next step could decide who survived it.

Chapter Nineteen
Doubt Creeps In

The morning sunlight spilled into the apartment, brushing across the floor in warm streaks. Lucy was humming while she stacked dishes in the kitchen, perfectly ordinary, perfectly human. Bea paused mid step, watching her. The fear she had carried for weeks suddenly felt misplaced. Lucy wasn't manipulating anything. She wasn't guiding them with hidden rules or subtle commands. She was normal.

Mina, Erin, Theo, and Bea gathered around the kitchen table, all noticeably tense. The radio in the corner crackled and sputtered, then played a soft melody that seemed to float around them like a warning:

"He waits beyond the world you know,
Keep your heart guarded, don't let him show,
Doors appear if you don't give in,
Perfect worlds fade if love slips in."

Bea's stomach twisted. The "he" in the lyrics suddenly seemed more real, more threatening than anything Lucy had ever done. She glanced at her friends, and they all looked uneasy, as if realizing the same thing at once.

"The crush," Mina whispered. "It's him. Not Lucy."

Erin nodded slowly. "All this time maybe Lucy was never dangerous. The songs are warnings about someone else. About not falling in love, not letting someone pull us into danger."

Bea shivered. Her hands shook slightly. "So everything we feared about Lucy maybe it was never true. She's safe. She's human. But he, he's the threat."

The radio hummed again, almost like a sigh, as if confirming their realization.

Chapter Twenty
Avoiding Temptation

The next few days became exercises in discipline. Every glance at the school hall, every passing smile, every faint thought of attraction was carefully monitored and suppressed. The friends didn't just avoid crushes. They trained themselves to ignore even the smallest sparks of interest.

Bea noticed herself watching her own thoughts, her feelings, her impulses. Every time she felt drawn to someone, even in the tiniest way, a sharp mental alarm went off. The songs repeated in her mind, instructions disguised as melodies:

"Eyes that linger, hearts that stray,
Doors appear to those who stay,
Trust the hand that gives no harm,
But keep away from the subtle charm."

"Step by step, doors are shown. Perfect worlds fade if love slips in," she whispered to herself. The lyrics pressed against her mind, a gentle but insistent warning she couldn't ignore.

Erin watched Bea carefully. "We have to follow the songs. Avoid the crush, ignore temptation. It's the only way."

Bea nodded, though her heart ached at the thought. They were safe from Lucy, but not safe from the rules, the songs, or the danger lurking just beyond the edges of the world they knew.

Lucy continued to act entirely normal. She baked muffins for everyone one morning, laughing at Theo's clumsy attempt to juggle his utensils. She helped Mina organize her notes and even teased Erin gently about spilling coffee the day before.

Bea observed her quietly, half expecting some hidden manipulation, some flicker of control. But there was nothing. Lucy was human. Ordinary. Innocent in the way she moved, spoke, and smiled.

For the first time in weeks, Bea felt a flicker of relief. But it didn't last. The world around them, the perfect coincidences, the keys appearing just when needed, the radio songs, still hummed with subtle insistence. Someone, or something, was orchestrating events. And it wasn't Lucy.

Mina glanced at Bea with a frown. "It's strange. We've spent so long fearing Lucy, but maybe we've been looking in the wrong place."

Bea nodded slowly. "Yes. The danger isn't Lucy. It's the crush, the rules, the Superworld itself. Lucy is part of our ordinary world. She's not controlling us."

Chapter Twenty-One
Following the Rules

The friends focused on the songs and the keys, treating them as a code to follow. They memorized lyrics, watched coincidences carefully, avoided any signs of temptation, and refrained from thinking about the crushes mentioned in the music.

"Hold your heart and guard your eyes,
The chosen paths are safe from lies,
Step by step, the door is shown,
But love misplaced will lead alone."

The rules were clear. Each key, each song, and each perfectly timed event was part of a pattern they couldn't yet fully understand, but they could follow it.

The tension in the group increased, though. Avoiding natural impulses and constantly monitoring their own feelings was exhausting. They all felt it: the safe path was narrow, and one misstep could unravel everything.

Bea held the second key tightly. She could feel the faint warmth in her palm, like it knew she was ready for the next step, the next door, but only if she followed the rules exactly.

Chapter Twenty-Two
A Glimpse Beyond

One evening, as the sun dipped behind the city, Bea noticed the air felt… different. The familiar scents, the bread, the coffee, even Lucy's perfume, shifted. Shadows grew longer, sharper at the edges, and the world seemed to stretch in subtle, unnatural ways.

The radio whispered another melody:

"Beyond the world you know,
Lies a place the chosen go,
Perfect days may fade to gray,
But doors you find may show the way."

Bea's stomach clenched. The Superworld was no longer distant. It was near, curling around their lives with invisible fingers, waiting for them to misstep.

Lucy, ordinary and harmless, laughed at a joke Mina told. She poured herself a cup of tea, completely unaware of the tension building around her. She was safe. Perfectly safe.

And that terrified Bea. Because if Lucy wasn't the danger… then the Superworld and the "crush" lurking beyond the edges of their lives were.

Bea realized with cold clarity that they were already approaching the edge of something vast, something beyond their understanding...

The day started like any other, but the air felt heavier, almost electric, as if the world itself was holding its breath. Bea clutched the second key in her pocket, her fingers tracing the carved symbols as she walked to meet Erin, Mina, and Theo.

"Do you feel it?" Bea asked quietly, her voice almost lost beneath the hum of the city.

Erin nodded, her eyes scanning the streets nervously. "Something's different. The patterns... the coincidences... they're sharper today. I think... we're close."

Bea's stomach twisted. She had been following the songs and keys for weeks, avoiding the "crush" and watching Lucy carefully. Now, for the first time, she felt as though the rules were converging toward a single point.

They arrived at an abandoned subway station, quiet and dimly lit. The radio played faintly from somewhere deep in the tunnels:

"Steps you take will show the way,
The chosen hearts must guard the day,
Keys unlock what eyes can't see,
Doors appear for those who flee."

Mina shivered. "This... this has to be it. One of the doors the songs keep talking about."

Bea nodded. The key grew warm in her hand. "Then we follow it. Step by step. No distractions. No mistakes."

The four of them stepped into the shadows, their own breaths loud and echoing in the tunnels. Every instinct screamed that they were leaving the ordinary world behind.

Chapter Twenty-Three
The Crush Lurks

As they descended into the subway tunnels, the faint echoes of the radio songs seemed to wrap around them. The lyrics repeated insistently in their minds:

"Eyes that linger, hearts that stray,
Step by step, doors guide the way,
Trust the hand that gives no harm,
But keep away from the subtle charm."

Bea felt her stomach twist. Every instinct told her that the "crush" was closer than ever, lurking somewhere just beyond the edge of the world they knew. The friends moved silently, each step measured, avoiding distractions and avoiding thoughts that could draw attention.

"Do you think… he's here?" Theo whispered.

Bea shook her head. "We don't know. But the songs warned us. That's why we can't look, can't care. We have to stay focused."

Mina's hands trembled slightly as she clutched her notebook, recording every observation. The stakes were rising. They had followed the songs and the keys this far, but the Superworld wasn't a distant threat anymore. It was waiting.

Chapter Twenty-Four
First Glimpse

A faint shimmer appeared in the shadows ahead, a doorway impossible in its geometry, glowing faintly with a light that felt both sweet and toxic. Bea's breath caught. This was the Superworld door.

The radio hummed again, the lyrics urging them forward:

"Beyond the world you know,
Lies the place the chosen go,
Perfect days may fade to gray,
But doors you find will show the way."

Erin's hand trembled as she reached for the doorknob. "It's real… it's actually real."

Bea nodded, heart pounding. The second key burned warmly in her pocket, as if urging her forward. Every step toward the door felt heavier, charged with unseen energy.

They crossed the threshold, and the ordinary world seemed to shiver, flicker, and fade. The air smelled different. Sweeter at first, like candy, but with an undertone of something… metallic, sharp, alive.

Bea realized with a cold jolt that the songs weren't just warnings. They were instructions. And if they had made a mistake, there was no undoing it.

The moment they stepped through, the world shifted. The sky was impossibly vibrant, colors bleeding into one another. The ground shimmered, soft and springy. Everything felt heightened, alive, and unreal.

Bea's stomach tightened. "This… this is the Superworld."

Theo swallowed nervously. "I thought it was... a metaphor. A warning. But it's real. We're really here."

Mina flipped through her notebook frantically. "The songs... they said the Superworld is temporary. That it could expire. We have to find the keys. We have to follow the rules exactly, or we... we don't get out."

Bea's eyes narrowed. "And the crush... whoever he is... we still can't let him get our attention. The rules haven't changed."

The radio played faintly, almost like a heartbeat beneath the surreal world:

"Guard your heart, step carefully,
Eyes may lure you deceptively,
Keys are your guide, the doors your fate,
Follow the path before too late."

Bea realized the true danger was now tangible. Every step could be a misstep. Every glance, every instinct, could betray them. The ordinary world, Lucy, the café, the streets, was gone. Only the Superworld and its rules remained.

Chapter Twenty-Five
Danger!

They moved carefully through the Superworld, each step calculated. The air shimmered with every sound they made, and the faint echoes of the songs guided them like a pulse. Keys appeared where the radio instructed, hidden under candy-colored bushes or behind impossible stone arches.

Bea noticed something terrifying: the Superworld itself seemed alive. It shifted subtly with their actions. Missteps made the colors flicker and shadows twitch unnaturally. The radio's voice grew more insistent:

"The chosen hearts must guard their path,
Avoid the eyes that spark the wrath,
Step by step, the door is shown,
But love misplaced will lead alone."

Bea's fingers tightened around the second key. "Step by step… avoid the crush… follow the doors… find the keys…"

Mina looked around nervously. "It's… like the Superworld is watching us. Testing us."

Bea nodded. "And it will punish us if we break the rules. If we… think about him, if we get distracted… if we don't follow the songs."

The friends moved in silence, hearts hammering. Each key they found was a small victory, each correct step a momentary relief. But Bea knew, deep down, that the Superworld was preparing them for something far worse.

The Superworld was alive. Colors shimmered unnaturally, candy-like scents twisted with a metallic tang, and shadows seemed to stretch longer than they should. Each step Bea, Erin, Mina, and Theo took felt like walking a tightrope.

"The chosen hearts must guard their way,
Do not stray or you will pay,
Keys will guide, doors will appear,
But hearts misplaced invite your fear."

The songs echoed faintly around them, repeating like a heartbeat. Every line was a warning. Every key they found glowed briefly when touched, pulsing with warmth.

Bea held the third key now, eyes wide. "The Superworld… it feels alive. Like it's… testing us."

Mina nodded, notebook clutched to her chest. "And the crush… the songs keep warning us. We can't think about anyone, can't give in. We have to stay perfect."

The air itself seemed to hum with tension. Something was coming. They could feel it, even though they didn't know what.

Chapter Twenty-Six
False Safety

Lucy was… still normal. She smiled, laughed, and poured tea. She wasn't guiding them. She wasn't dangerous. She was human, ordinary, comforting in a way that made Bea's stomach twist.

"Maybe we were wrong all along," Erin whispered as they navigated a glowing bridge of impossible shapes. "Lucy… she's safe. It's not her."

Bea nodded, though unease still gnawed at her. "Yes… but the Superworld itself isn't safe. And the songs… they haven't lied yet. If we slip… it will punish us."

The crush became a shadow in their minds, never seen, but always felt. Every instinct, every thought, every attraction had to be controlled. Step by step, key by key, they advanced through the surreal landscape.

The closer they got to the exit, the heavier the air became. It smelled less like candy and more like… something burning, something real.

By the time the fourth key appeared, the friends were exhausted. Their hands shook, their hearts raced, but they had memorized every song lyric, every instruction, every hidden warning.

Bea held it carefully, warmth pulsing like a living thing. The radio hummed its most urgent song yet:

"Guard your hearts, your paths are near,
The chosen doors are always clear,
But love misled will fade away,
The final key will show the way."

"This has to be it," Mina whispered. "The last key. The exit."

Bea swallowed. Every instinct screamed that they were close. Every shadow, every shimmer of the Superworld, warned of impending danger.

Chapter Twenty-Seven
The Crush's Shadow

The "crush" from the songs was never seen, never named, but his presence was like a current beneath the air. Every distraction, every misstep could draw attention and collapse their path.

Bea glanced at her friends. "We've followed the rules… avoided him… stayed focused."

Theo nodded, his voice shaking. "But what if… he's closer than we think?"

Mina shook her head. "It doesn't matter. The songs… the keys… they told us what to do. That's the only way to survive."

And just as the final doorway appeared, shaped like a star and glowing faintly, the Superworld seemed to pulse, threatening collapse. They could feel the expiration ticking down, something primal and terrifying that no song or key could fully describe.

Chapter Twenty-Eight
The Final Rush

A came rushing in, sweat beading on her forehead, breathing ragged, eyes wide. She grabbed Bea frantically.

"B! B! The song, they weren't about the crush! We… we already secured our spot…"

Bea's heart lurched. "What do you mean?! The key, where is it?!"

"The songs! They were about Lucy! Lucy brought us in! The Superworld… it's her!"

Before Bea could respond, the Superworld shimmered violently. Colors twisted, the air burned, and shadows warped. The star-shaped key glowed, then warped in her hands.

Bea screamed, desperate, running toward the exit. She shoved the key into the doorway.

It melted. Bent. Gone.

Lucy's voice echoed softly, almost sweetly, from down the glowing corridor.

"So you know the songs were about me. How thoughtful of you all."

Bea's chest tightened. "You… YOU did this! This isn't real..."

Lucy's smile was calm. "It's real enough. You were miserable, helpless. I gave your life meaning… and the Superworld… I control it. Every door, every key, every song. And I think I'm getting tired of yours already."

The colors flared. The air shifted.

The Superworld began to collapse. Candy-like air turned acrid, shadows jagged, light flickering. Every step became harder, as if gravity itself was unstable.

Bea ran, her mind racing, but the key in her hands was gone. The exit glimmered, but the doorway flickered and warped. She could hear Lucy's soft laughter echoing behind her, echoing in every direction.

"It has to happen. Your lives… your choices… all led here. And now, it's ending."

Bea's friends shouted, ran, and grabbed her hand, but the Superworld was alive, collapsing, reshaping around them. The expiration reached its climax.

Chapter Twenty-Nine
Gone

Bea's heart pounded as she shoved forward toward the exit. Colors twisted violently, and the scent of sugar was replaced by burnt air.

She glanced back. A's face was gone, melted into the collapsing world. Erin, Mina, and Theo clung to her, but the doors flickered, shimmering out of reach.

Lucy's figure appeared faintly down the hall, calm, smiling, almost ethereal.

"You should be thanking me. You wouldn't have survived in the ordinary world without me. But your time here… is over."

Bea screamed, but the exit dissolved in her hands. The key melted into nothing. The Superworld… expired.

Everything went white.

Bea didn't get out. The Superworld collapsed. Lucy's control had been absolute. The songs, the keys, and the careful avoidance of the crush were all part of a test, a trap, a world that ended because it was never meant to last.

The perfect lives, the Superworld, the warnings, they were gone.

And Lucy… still smiled.

Made in the USA
Las Vegas, NV
21 March 2026

44048273R00036